GRETA THUNBERG

NO ONE
IS TOO SMALL
TO MAKE
A DIFFERENCE

allen lane

an imprint of
PENGUIN BOOKS

GRETA THUNBERG was born in 2003. In August 2018, she decided not to go to school one day, starting a strike for the climate outside the Swedish Parliament. Her actions ended up sparking a global movement for action against the climate crisis, inspiring millions of pupils to go on strike for our planet, and earning her the prestigious Prix Liberté, as well as a Nobel Peace Prize nomination. Greta has Asperger's, and considers it a gift which has enabled her to see the climate crisis 'in black and white'.

No One Is Too Small to Make a Difference is Greta's first book in English, collecting her speeches from climate rallies across Europe to audiences at the UN, the World Economic Forum, and the British Parliament. Her next book, *Our House is on Fire: Scenes of a Family and a Planet in Crisis*, is a memoir, jointly written with her mother, the opera singer Malena Ernman, her sister Beata Ernman, and her father Svante Thunberg.

Contents

September 2018

Stockholm

Every single person counts.

Our school strike has nothing to do with party politics.

Climate March
Stockholm, Sweden
8 September 2018

Our Lives are in Your Hands

Last summer, a number of leading climate scientists wrote that we have at most three years to reverse growth in greenhouse-gas emissions if we're going to reach the goals set in the Paris Agreement.

Over a year and two months have now passed, and in that time many other scientists have said the same thing and a lot of things have got worse and greenhouse-gas emissions continue to increase. So maybe we have even less time than the one year and ten months those scientists said we have left.

If people knew this they wouldn't need to ask me why I'm so 'passionate about climate change'.

If people knew that the scientists say that we have a 5 per cent chance of meeting the Paris target, and if people knew what a nightmare scenario we will face if we don't keep global warming below 2°C, they wouldn't need to ask me why I'm on school strike outside parliament.

Because if everyone knew how serious the situation is and how little is actually being done, everyone would come and sit down beside us.

In Sweden, we live our lives as if we had the resources of 4.2 planets. Our carbon footprint is one of the worst in the world. This means that Sweden steals 3.2 years of natural resources from future generations every year. Those of us who are part of these future generations would like Sweden to stop doing that.

Right now.

This is not a political text. Our school strike has nothing to do with party politics.

Because the climate and the biosphere don't care about our politics and our empty words for a single second.

They only care about what we actually do.

This is a cry for help.

To all the newspapers who still don't write about and report on climate change, even though they said that the climate was 'the critical question of our time' when the Swedish forests were burning this summer.

To all of you who have never treated this crisis as a crisis.

To all the influencers who stand up for everything except the climate and the environment.

To all the political parties that pretend to take the climate question seriously.

To all the politicians that ridicule us on social media, and have named and shamed me so that people tell me that I'm retarded, a bitch and a terrorist, and many other things.

To all of you who choose to look the other way every day because you seem more frightened of the changes that can prevent catastrophic climate change than the catastrophic climate change itself.

Your silence is almost worst of all.

The future of all the coming generations rests on your shoulders.

Those of us who are still children can't change what you do now once we're old enough to do something about it.

A lot of people say that Sweden is a small country, that it doesn't matter what we do. But I think that if a few girls can get headlines all over the world just by not going to school for a few weeks, imagine what we could do together if we wanted to.

Every single person counts.

Just like every single emission counts.

Every single kilo.

Everything counts.

So please, treat the climate crisis like the acute crisis it is and give us a future.

Our lives are in your hands.

October
2018

Helsinki

London

Almost Everything is Black and White

When I was about eight years old, I first heard about something called climate change, or global warming. Apparently, that was something humans had created by our way of living. I was told to turn off the lights to save energy, and to recycle paper to save resources.

I remember thinking that it was very strange that humans, who are an animal species among others, could be capable of changing the earth's climate.

Because, if we were and if it was really happening, we wouldn't be talking about anything else. As soon as you turned on the TV, everything would be about that. Headlines, radio, newspapers. You would never read or hear about anything else. As if there was a world war going on.

But. No one talked about it. Ever.

If burning fossil fuels was so bad that it threatened

our very existence, how could we just continue like before? Why were there no restrictions? Why wasn't it made illegal?

To me, that did not add up. It was too unreal.

I have Asperger's syndrome, and to me, almost everything is black or white.

I think in many ways that we autistic are the normal ones and the rest of the people are pretty strange. They keep saying that climate change is an existential threat and the most important issue of all. And yet they just carry on like before. If the emissions have to stop, then we must stop the emissions. To me that is black or white. There are no grey areas when it comes to survival. Either we go on as a civilization or we don't.

We have to change.

Countries like Sweden and the UK need to start reducing emissions by at least 15 per cent every year, to stay below a 2°C warming target.

But, as the IPCC has recently stated, aiming instead for a 1.5°C target would significantly reduce the climate impact. But we can only imagine what that means for reducing emissions. You would think every one of our leaders and the media would be talking about nothing else – but no one ever mentions it. Nor does anyone ever mention anything about the greenhouse gases already locked in the system, nor that air pollution is hiding a warming, so when we stop burning fossil fuels, we already have an extra 0.5–1.1°C guaranteed.

Nor does hardly anyone ever mention that we are in the midst of the sixth mass extinction, with about 200 species going extinct every single day.

Furthermore, does no one ever speak about the aspect of equity, or climate justice, clearly stated everywhere in the Paris Agreement and the Kyoto Protocol, which is absolutely necessary to make the Paris Agreement work on a global scale? That means that rich countries need to get down to zero emissions, within six to twelve years, so that people in poorer countries can heighten their standard of living by building some of the infrastructure that we have already built. Such as roads, hospitals, electricity, schools and clean drinking water. Because how can we expect countries like India or Nigeria to care about the climate crisis if we, who already have everything, don't care even a second about it or our actual commitments to the Paris Agreement?

So, why are we not reducing our emissions? Why are they, in fact, still increasing? Are we knowingly causing a mass extinction? Are we evil?

No, of course not. People keep doing what they do because the vast majority doesn't have a clue about the consequences of our everyday life. And they don't know the rapid changes required.

Since, as I said before, no one talks about it. There are no headlines, no emergency meetings, no breaking news. No one is acting as if we were in a crisis. Even most green politicians and climate scientists go on flying around the world, eating meat and dairy.

If I live to be 100 I will be alive in the year 2103.

When you think about 'the future' today, you don't think beyond the year 2050. By then I will, in the best case, not even have lived half of my life. What happens next?

The year 2078 I will celebrate my seventy-fifth birthday.

What we do or don't do, right now, will affect my entire life, and the lives of my children and grandchildren.

When school started in August this year I decided that this was enough. I sat myself down on the ground outside the Swedish parliament. I school-striked for the climate.

Some people say that I should be in school instead.

Some people say that I should study to become a climate scientist so that I can 'solve the climate crisis'. But the climate crisis has already been solved.

We already have all the facts and solutions. All we have to do is to wake up and change.

And why should I be studying for a future that soon will be no more, when no one is doing anything whatsoever to save that future?

And what is the point of learning facts within the school system when the most important facts given by the finest science of that same school system clearly mean nothing to our politicians and our society?

A lot of people say that Sweden is just a small country, and that it doesn't matter what we do. But I think that if a few children can get headlines all over the world just by not going to school for a few weeks, imagine what we all could do together if we wanted to.

Today we use 100 million barrels of oil every day.

There are no politics to change that. There are no rules to keep that oil in the ground.

So we can't save the world by playing by the rules.

Because the rules have to be changed.

Everything needs to change. And it has to start today.

So everyone out there: it is now time for civil disobedience.

It is time to rebel.

We can't save the world by playing by the rules.

December
2018

Stockholm

 COP24·KATOWICE
UNITED NATIONS CLIMATE CHANGE CONFERENCE
POLAND 2018

Unpopular

My name is Greta Thunberg, I am fifteen years old and I'm from Sweden. I speak on behalf of Climate Justice Now.

Many people say that Sweden is just a small country and it doesn't matter what we do. But I've learnt that no one is too small to make a difference. And if a few children can get headlines all over the world just by not going to school – then imagine what we all could do together if we really wanted to.

But to do that we have to speak clearly. No matter how uncomfortable that may be. You only speak of green, eternal economic growth because you are too scared of being unpopular. You only talk about moving forward with the same bad ideas that got us into this mess. Even when the only sensible thing to do is to pull the emergency brake.

You are not mature enough to tell it like it is. Even that burden you leave to your children. But I don't care about being popular, I care about climate justice and the living planet.

We are about to sacrifice our civilization for the opportunity of a very small number of people to continue to make enormous amounts of money. We are about to sacrifice the biosphere so that rich people in countries like mine can live in luxury. But it is the sufferings of the many which pay for the luxuries of the few.

The year 2078 I will celebrate my seventy-fifth birthday.

If I have children, then maybe they will spend that day with me. Maybe they will ask about you.

Maybe they will ask why you didn't do anything, while there still was time to act. You say that you love your children above everything else. And yet you are stealing their future.

Until you start focusing on what needs to be done rather than what is politically possible, there's no hope. We cannot solve a crisis without treating it as a crisis. We need to keep the fossil fuels in the ground and we need to focus on equity.

And if solutions within this system are so impossible to find then maybe we should change the system itself?

We have not come here to beg world leaders to care. You have ignored us in the past and you will ignore us again. You've run out of excuses and we're running out of time. We've come here to let you know that change is coming whether you like it or not.

The real power belongs to the people.

We are about to sacrifice our civilization for the opportunity of a very small number of people to continue to make enormous amounts of money.

January
2019

Stockholm

Prove Me Wrong

Some people say that we are not doing enough to fight climate change. But that is not true. Because to 'not do enough' you have to do something. And the truth is we are basically not doing anything.

Yes, some people are doing more than they can but they are too few or too far away from power to make a difference today.

Some people say that the climate crisis is something that we all have created. But that is just another convenient lie. Because if everyone is guilty then no one is to blame.

And someone is to blame. Some people – some companies and some decision-makers in particular – have known exactly what priceless values they are sacrificing to continue making unimaginable amounts of money.

I want to challenge those companies and those decision-makers into real and bold climate action. To set their economic goals aside and to safeguard the future living conditions for humankind. I don't believe for one second that you will rise to that challenge. But I want to ask you all the same.

I ask you to prove me wrong. For the sake of your children, for the sake of your grandchildren. For the sake of life and this beautiful living planet.

I ask you to stand on the right side of history. I ask you to pledge to do everything in your power to push your own business or government in line with a 1.5°C world.

Will you pledge to do that? Will you pledge to join me, and the people all around the world, in doing whatever it takes?

I ask you to stand on the right side of history.

World Economic Forum
Davos, Switzerland
25 January 2019

Our House is On Fire

Our house is on fire.

I am here to say, our house is on fire.

According to the IPCC, we are less than twelve years away from not being able to undo our mistakes.

In that time, unprecedented changes in all aspects of society need to have taken place – including a reduction of our CO_2 emissions by at least 50 per cent.

And please note that those numbers do not include the aspect of equity, which is absolutely necessary to make the Paris Agreement work on a global scale.

Nor does it include tipping points or feedback loops like the extremely powerful methane gas released from the thawing Arctic permafrost.

At places like Davos, people like to tell success stories. But their financial success has come with an unthinkable price-tag. And on climate change, we have to acknowledge that we have failed.

All political movements in their present form have done so.

And the media has failed to create broad public awareness.

But *Homo sapiens* have not yet failed. Yes, we are failing, but there is still time to turn everything around. We can still fix this. We still have everything in our own hands.

But unless we recognize the overall failures of our current systems we most probably don't stand a chance.

We are facing a disaster of unspoken sufferings for enormous amounts of people. And now is not the time for speaking politely or focusing on what we can or cannot say. Now is the time to speak clearly.

Solving the climate crisis is the greatest and most complex challenge that *Homo sapiens* have ever faced. The main solution, however, is so simple that even a small child can understand it.

We have to stop our emissions of greenhouse gases.

And either we do that or we don't.

You say that nothing in life is black or white.

But that is a lie. A very dangerous lie.

Either we prevent a 1.5°C of warming or we don't.

Either we avoid setting off that irreversible chain reaction beyond human control – or we don't.

Either we choose to go on as a civilization or we don't.

That is as black or white as it gets.

There are no grey areas when it comes to survival.

Now we all have a choice.

We can create transformational action that will safeguard the living conditions for future generations.

Or we can continue with our business as usual and fail.

That is up to you and me.

Some say that we should not engage in activism.

Instead we should leave everything to our politicians and just vote for a change instead. But what do we do when there is no political will? What do we do when the politics needed are nowhere in sight?

Here in Davos – just like everywhere else – everyone is talking about money. It seems that money and growth are our only main concerns.

And since the climate crisis is a crisis that never once has been treated as a crisis, people are simply not aware of the full consequences from our everyday life. People are not aware that there is such a thing as a carbon budget and just how incredibly small that remaining carbon budget is. And that needs to change today.

No other current challenge can match the importance of establishing a wide, public awareness and understanding of our rapidly disappearing carbon budget, that should and must become our new global currency and the very heart of our future and present economics.

We are now at a time in history where everyone

with any insight of the climate crisis that threatens our civilization and the entire biosphere must speak out.

In clear language.

No matter how uncomfortable and unprofitable that may be.

We must change almost everything in our current societies.

The bigger your carbon footprint – the bigger your moral duty.

The bigger your platform – the bigger your responsibility.

Adults keep saying: 'We owe it to the young people to give them hope.'

But I don't want your hope.

I don't want you to be hopeful.

I want you to panic.

I want you to feel the fear I feel every day.

And then I want you to act.

I want you to act as you would in a crisis.

I want you to act as if our house is on fire.

Because it is.

The bigger your carbon footprint – the bigger your moral duty.

February
2019

Stockholm

SPÅR 11 · 14:25 · Tågnr 537 · Bistro
Norrköping Köpenhamn H
Linköping Alvesta Malmö
Vagnsordning 1, 3, 4, 5, 6, 7

SKOLSTREJK FÖR KLIMATET

Facebook
Stockholm, Sweden
2 February 2019

I'm Too Young to Do This

Recently I've seen many rumours circulating about me and enormous amounts of hate. This is no surprise to me. I know that since most people are not aware of the full meaning of the climate crisis (which is understandable since it has never been treated as a crisis) a school strike for the climate would seem very strange to people in general. So let me make some things clear about my school strike.

In May 2018 I was one of the winners in a writing competition about the environment held by *Svenska Dagbladet*, a Swedish newspaper. I got my article published and some people contacted me, among others was Bo Thorén from Fossil Free Dalsland. He had some kind of group with people, especially youth, who wanted to do something about the climate crisis.

I had a few phone meetings with other activists.

The purpose was to come up with ideas of new projects that would bring attention to the climate crisis. Bo had a few ideas of things we could do. Everything from marches to a loose idea of some kind of school strike (that schoolchildren would do something on the schoolyards or in the classrooms). That idea was inspired by the Parkland students, who had refused to go to school after the school shootings.

I liked the idea of a school strike. So I developed that idea and tried to get the other young people to join me, but no one was really interested. They thought that a Swedish version of the Zero Hour march was going to have a bigger impact. So I went on planning the school strike all by myself and after that I didn't participate in any more meetings.

When I told my parents about my plans, they weren't very fond of it. They did not support the idea of school striking and they said that if I were to do this I would have to do it completely by myself and with no support from them.

On the 20th of August I sat down outside the Swedish parliament. I handed out fliers with a long list of facts about the climate crisis and explanations on why I was striking. The first thing I did was to post on Twitter and Instagram what I was doing and it soon went viral. Then journalists and newspapers started to come. A Swedish entrepreneur and businessman active in the climate movement, Ingmar Rentzhog, was among the first to arrive. He spoke with me and took pictures that he posted on Facebook. That was the first time I had ever met or spoken with him. I had not communicated or encountered with him ever before.

Many people love to spread rumours saying that I have people 'behind me' or that I'm being 'paid' or 'used' to do what I'm doing. But there is no one 'behind' me except for myself. My parents were as far from climate activists as possible before I made them aware of the situation.

I am not part of any organization. I sometimes supported and cooperated with several NGOs that work with the climate and environment. But I am absolutely independent and I only represent myself. And I do what I do completely for free, I have not received any money or any promise of future payments in any form at all. And nor has anyone linked to me or my family done so.

And of course it will stay this way. I have not met one single climate activist who is fighting for the climate for money. That idea is completely absurd.

Furthermore, I only travel with permission from my school, and my parents pay for tickets and accommodation.

My family has written a book together about our family and how I and my sister, Beata, have influenced my parents' way of thinking and seeing the world, especially when it comes to the climate. And about our diagnoses.

That book was due to be released in May 2018. But since there was a major disagreement with the book company, we ended up changing to a new publisher, and so the book was released in August the same year instead.

Before the book was released my parents made it clear that their possible profits from the book, *Scener ur hjärtat*, 'Scenes From the Heart', will be going to eight different charities working with the environment, children with diagnoses and animal rights.

And yes, I write my own speeches. But since I know that what I say is going to reach many, many people, I often ask for input. I also have a few scientists that I frequently ask for help on how to express certain complicated matters. I want everything to be absolutely correct so that I don't spread incorrect facts, or things that can be misunderstood.

Some people mock me for my diagnosis. But Asperger is not a disease, it's a gift. People also say that since I have Asperger I couldn't possibly have put myself in this position. But that's exactly why I did this. Because if I would have been 'normal' and social I would have organized myself in an organization, or started an organization by myself.

But since I am not that good at socializing I did this instead. I was so frustrated that nothing was being done about the climate crisis, and I felt like I had to do something, anything. And sometimes NOT doing things – like just sitting down outside the parliament – speaks much louder than doing things. Just like a whisper sometimes is louder than shouting.

Also there is one complaint that I 'sound and write

like an adult'. And to that I can only say: Don't you think that a sixteen-year-old can speak for herself?

There are some people who say that I oversimplify things. For example when I say that 'The climate crisis is a black and white issue', 'We need to stop the emissions of greenhouse gases', and 'I want you to panic.' But that I only say because it's true. Yes, the climate crisis is the most complex issue that we have ever faced and it's going to take everything from our part to 'stop it'. But the solution is black and white: we need to stop the emissions of greenhouse gases.

Because either we limit the warming to 1.5°C over pre-industrial levels, or we don't. Either we reach a tipping point where we start a chain reaction with events way beyond human control, or we don't. Either we go on as a civilization, or we don't. There are no grey areas when it comes to survival.

And when I say that I want you to panic, I mean that we need to treat the crisis as a crisis. When your house is on fire you don't sit down and talk about how nice you can rebuild it once you put out the fire. If your house is on fire you run outside and make sure that everyone is out while you call the fire department. That requires some level of panic.

There is one other argument that I can't do anything about. And that is the fact that I'm 'just a child and we shouldn't be listening to children'. But that is easily fixed – just start to listen to the rock-solid science instead. Because if everyone listened to the scientists and the facts that I constantly refer to then no one would have to listen to me or any of the other hundreds of thousands of schoolchildren on strike for the climate across the world.

Then we could all go back to school. I am just a messenger, and yet I get all this hate. I am not saying anything new, I am just saying what scientists have repeatedly said for decades.

And I agree with you, I'm too young to do this.

We children shouldn't have to do this. But since almost no one is doing anything, and our very future is at risk, we feel like we have to continue.

And if you have any other concern or doubt about me, then you can listen to my TED talk, in which I talk about how my interest for the climate and environment began.

And thank you everyone for your kind support! It brings me hope.

We children shouldn't have to do this.

I'm Too Young to Do This

SAVE OUR FUTURE

AKE UR NET AT AIN

NOTRE PLANTE
NOTRE CHOIX
NOTRE PLANTE

PUT YOUR GLASSES ON AND FACE

in a without

Brussels

You're Acting Like Spoiled, Irresponsible Children

My name is Greta Thunberg, I am a climate activist from Sweden and today in this room there are also – if you can come up – Anuna, Adélaïde, Kyra, Gilles, Dries, Toon and Luisa.

Tens of thousands of children or schools are striking for the climate on the streets of Brussels. Hundreds of thousands are doing the same all over the world. We are school-striking because we have done our homework. And some of us are here today. People always tell us that they are so hopeful. They are hopeful that the young people are going to save the world, but we are not. There is simply not enough time to wait for us to grow up and become the ones in charge. Because by the year 2020 we need to have bended the emissions curve steep downward.

That is next year. We know that most politicians

don't want to talk to us. Good, we don't want to talk to them either. We want them to talk to the scientists instead. Listen to them, because we are just repeating what they are saying and have been saying for decades. We want you to follow the Paris Agreement and the IPCC reports. We don't have any other manifestos or demands – you unite behind the science, that is our demand. When many politicians talk about the school strike for the climate, they talk about almost anything except for the climate crisis.

Many people are trying to make the school strikes a question of whether we are promoting truancy or whether we should go back to school or not. They make up all sorts of conspiracies and call us puppets who cannot think for ourselves. They are desperately trying to remove the focus from the climate crisis and change the subject. They don't want to talk about it because they know they cannot win this fight. Because they know they haven't done their homework, but we have. Once you have done your homework you realize that we need new politics, we need new economics where everything is based on a rapidly declining and extremely limited remaining carbon budget.

But that is not enough. We need a whole new way of thinking. The political system that you have created is all about competition. You cheat when you can, because all that matters is to win, to get power. That must come to an end, we must stop competing with each other, we need to cooperate and work together and to share the resources of the planet in a fair way.

We need to start living within the planetary boundaries, focus on equity and take a few steps back for the sake of all living species. We need to protect the biosphere, the air, the oceans, the soil, the forests.

This may sound very naive, but if you have done your homework then you know that we don't have any other choice. We need to focus every inch of our being on climate change, because if we fail to do so then all our achievements and progress have been for nothing and all that will remain of our political leaders' legacy will be the greatest failure of human history. And they will be remembered as the greatest villains of all time, because they have chosen not to listen and not to act. But this does not have to be. There is still time. According to the IPCC report we are about eleven years away from being in a position where we set off an irreversible chain reaction beyond human control.

To avoid that unprecedented change in all aspects of society, [actions] need to have taken place within this coming decade, including a reduction of our CO_2 emissions by at least 50 per cent by the year 2030. And please note that those numbers do not include the aspect of equity, which is absolutely necessary to make the Paris Agreement work on a global scale, nor do they include tipping points or feedback loops like the extremely powerful methane gas released from the thawing Arctic permafrost.

They do, however, include negative emission techniques on a huge planetary scale that is yet to be invented, and that many scientists fear will never be ready in time and will anyway be impossible to deliver at the scale assumed. We have been told that the EU intends to improve its emission reduction targets. In the new target, the EU is proposing to reduce its greenhouse-gas emissions to 45 per cent below 1990's level by 2030. Some people say that is good or that is ambitious. But this new target is still not enough to keep global warming below 1.5°C.

This target is not sufficient to protect the future for children growing up today. If the EU is to make its fair contribution to staying within the carbon budget for the 2°C limit, then it means a minimum of 80 per cent reduction by 2030 and that includes aviation and shipping. So, it is around twice as ambitious as the current proposal. The actions required are beyond manifestos or any party politics. Once again, they sweep their mess under the carpet for our generation to clean up and solve. Some people say that we are fighting for our future, but that is not true. We are not fighting for our future, we are fighting for everyone's future. And if you think that we should be in school instead, then we suggest that you take our place in the streets striking from your work. Or better yet, join us so it can speed up the process.

And I am sorry, but saying everything will be all right while continuing doing nothing at all is just not hopeful to us. In fact, it's the opposite of hope. And yet this is exactly what you keep doing. You can't just sit around waiting for hope to come – you're acting like spoiled, irresponsible children. You don't seem to understand that hope is something you have to earn. And if you still say that we are wasting valuable lesson time, then let me remind you that our political leaders have wasted decades through denial and inaction. And since our time is running out we have decided to take action. We have started to clean up your mess and we will not stop until we are done.

Most politicians don't want to talk to us. Good, we don't want to talk to them either. We want them to talk to the scientists instead.

March
2019

A Strange World

I dedicate this award to the people fighting to protect the Hambach Forest. And to activists everywhere who are fighting to keep the fossil fuels in the ground.

We live in a strange world, where all the united science tells us that we are about eleven years away from setting off an irreversible chain reaction, way beyond human control, that will probably be the end of our civilization as we know it.

We live in a strange world, where children must sacrifice their own education in order to protest against the destruction of their future.

Where the people who have contributed the least to this crisis are the ones who are going to be affected the most.

Where politicians say it's too expensive to save the world, while spending trillions of euros subsidizing fossil fuels.

We live in a strange world, where no one dares to look beyond our current political systems even though it's clear that the answers we seek will not be found within the politics of today.

Where some people seem to be more concerned about the presence in school of some children than the future of humankind.

Where everyone can choose their own reality and buy their own truth.

Where our survival is depending on a small, rapidly disappearing carbon budget. And hardly anyone even knows it exists.

We live in a strange world, where we think we can buy or build our way out of a crisis that has been created by buying and building things.

Where a football game or a film gala gets more media attention than the biggest crisis humanity has ever faced.

Where celebrities, film and pop stars who have stood up against all injustices will not stand up for our environment and for climate justice because that would inflict on their right to fly around the world visiting their favourite restaurants, beaches and yoga retreats.

Avoiding catastrophic climate breakdown is to do the seemingly impossible. And that is what we have to do.

But here is the truth: we can't do it without you in the audience here tonight.

People see you celebrities as gods. You influence billions of people. We need you.

You can use your voice to raise awareness about this global crisis. You can help turn individuals into movements. You can help us wake up our leaders – and let them know that our house is on fire.

We live in a strange world.

But it's the world that my generation has been handed. It's the only world we've got.

We are now standing at a crossroads in history.

We are failing but we have not yet failed.

We can still fix this.

It's up to us.

A football game or a film gala gets more media attention than the biggest crisis humanity has ever faced.

**April
2019**

KOLSTREJK
FÖR
KLIMATET

Stockholm

Cathedral Thinking

My name is Greta Thunberg. I am sixteen years old. I come from Sweden. And I want you to panic.

I want you to act as if your house was on fire.

I have said those words before.

A lot of people have explained why that is a bad idea. A great number of politicians have told me that panic never leads to anything good.

And I agree. To panic unless you have to is a terrible idea. But when your house is on fire and you want to keep your house from burning to the ground then that does require some level of panic.

Our civilization is so fragile it is almost like a castle built in the sand. The façade is so beautiful but the foundations are far from solid.

We have been cutting so many corners.

Yesterday the world watched with despair and

enormous sorrow how Notre-Dame burned in Paris. Some buildings are more than just buildings. But Notre-Dame will be rebuilt. I hope that its foundations are strong. I hope that our foundations are even stronger. But I fear they are not.

Around the year 2030, 10 years, 259 days and 10 hours away from now, we will be in a position where we will set off an irreversible chain reaction beyond human control that will most likely lead to the end of our civilization as we know it. That is, unless in that time permanent and unprecedented changes in all aspects of society have taken place, including a reduction of our CO_2 emissions by at least 50 per cent. And please note that these calculations are depending on inventions that have not yet been invented at scale. Inventions that are supposed to clear our atmosphere of astronomical amounts of carbon dioxide.

Furthermore these calculations do not include unforeseen tipping points and feedback loops, like the extremely powerful methane gas escaping from rapidly thawing Arctic permafrosts. Nor do they include already locked-in warming, hidden by air pollution, nor the aspect of equity, nor climate justice, clearly stated throughout the Paris Agreement, which is absolutely necessary to make it work on a global scale.

We must also bear in mind that these are just calculations, estimations, meaning that the point of no return may occur a bit sooner or later than that. No one can know for sure. We can however be certain that they will occur approximately in these time-frames. Because these calculations are not opinions or wild guesses. These predictions are backed up by scientific facts, concluded by all nations through the IPCC.

Nearly every major scientific body around the world unreservedly supports the work and findings of the IPCC.

We are in the midst of the sixth mass extinction and the extinction rate is up to 10,000 times faster than what is considered normal, with up to 200 species becoming extinct every single day.

Erosion of fertile top soil. Deforestation of our great forests. Toxic air pollution. Loss of insects and wildlife. The acidification of our oceans.

These are all disastrous trends being accelerated by a way of life that we, here in our financially fortunate part of the world, see as our right to simply carry on.

But hardly anyone knows about these catastrophes or understands that they are just the first few symptoms of climate ecological breakdown.

Because how could they? They have not been told. Or more importantly, they have not been told by the right people, and in the right way.

Our house is falling apart.

Our leaders need to start acting accordingly.

Because at the moment they are not.

If our house was falling apart our leaders wouldn't go on like you do today. You would change almost every part of your behaviour as you do in an emergency. If our house was falling apart, you wouldn't fly around the world in business class, chatting about how the market will solve everything with clever, small solutions to specific, isolated problems.

You wouldn't talk about buying and building your way out of a crisis that has been created by buying and building things.

If our house was falling apart, you wouldn't hold three emergency Brexit summits, and no emergency summit regarding the breakdown of the climate and ecosystems.

You wouldn't be arguing about phasing out coal in fifteen or eleven years. If our house was falling apart, you wouldn't be celebrating that one single nation, like Ireland, may soon divest from fossil fuels.

You wouldn't celebrate that Norway has decided to stop drilling for oil outside the scenic resort of the Lofoten Islands, but will continue to drill for oil everywhere else, for decades.

It's thirty years too late for that kind of celebration.

If our house was falling apart the media wouldn't be writing about anything else.

The ongoing climate and ecological crisis would make up all the headlines.

If our house was falling apart, you wouldn't say that you have the situation under control, and place the future living conditions of all species in the hands of inventions that are yet to be invented. And you would not spend all your time, as politicians, arguing about taxes or Brexit.

If the walls of our house truly came tumbling town, surely you would set your differences aside and start cooperating.

Well, our house is falling apart. And we are rapidly running out of time. And yet basically nothing is happening.

Everyone and everything needs to change. So why waste precious time arguing about what and who needs to change first?

Everyone and everything has to change. But the bigger your platform, the bigger your responsibility. The bigger your carbon footprint, the bigger your moral duty.

When I tell politicians to act now, the most common answer is that they can't do anything drastic because it would be too unpopular among the voters.

And they are right, of course, since most people are not even aware of why those changes are required. That is why I keep telling you to unite behind the science. Make the best available science the heart of politics and democracy.

The EU elections are coming up soon and many of us who will be affected the most by this crisis, people like me, are not allowed to vote. Nor are we in a position to shape the decisions of business, politics, engineering, media, education or science. Because the time it takes for us to educate ourselves to do that simply does no longer exist. And that is why millions of children are taking to the streets, school-striking for the climate to create attention for the climate crisis.

You need to listen to us, we who cannot vote.

You need to vote for us, for your children and grandchildren.

What we are doing now can soon no longer be undone.

In this election, you vote for the future living conditions of humankind.

And though the politics needed do not exist today, some alternatives are certainly less worse than others. And I have read in the newspapers that some parties do not even want me standing here today because they so desperately do not want to talk about climate breakdown.

Our house is falling apart. The future as well as what we have achieved in the past is literally in your hands now. But it is still not too late to act. It will take a far-reaching vision. It will take courage. It will take fierce determination to act now, to lay the foundations when we may not know all the details about how to shape the ceiling.

In other words, it will take cathedral thinking.

I ask you to please wake up and make the changes required possible.

To do your best is no longer good enough.

We must all do the seemingly impossible.

And it's okay if you refuse to listen to me. I am after all just a sixteen-year-old schoolgirl from Sweden. But you cannot ignore the scientists, or the science, or the millions of schoolchildren who are school-striking for their right to a future.

I beg you, please do not fail in this.

———————

I have read in the newspapers that some parties do not even want me standing here today because they so desperately do not want to talk about climate breakdown.

Together We are Making a Difference

It's an honour for me to be here with you today. Together we are making a difference.

I come from Sweden, and back there it's almost the same problem as here, as everywhere, that nothing is being done to stop the climate and ecological crisis, despite all the beautiful words and promises.

We are now facing an existential crisis – the climate crisis, the ecological crisis – which has never been treated as a crisis before.

They have been ignored for decades. And for way too long the politicians and the people in power have gotten away with not doing anything at all to fight the climate crisis and the ecological crisis. But we will make sure that they do not get away with it any longer.

Humanity is now standing at a crossroads. We must now decide which path we want to take. What do we want the future living conditions for all species to be like?

We have gathered here today and in many other places around London and the world, because we have chosen the path we want to take, and now we are waiting for others to follow our example.

We are the ones making a difference. We, the people in Extinction Rebellion, and those school striking for the climate, we are making a difference. It shouldn't be like that, but since no one else is doing anything, we will have to do so.

And we will never stop fighting, we will never stop fighting for this planet, and for ourselves, our futures, and for the futures of our children and our grand-children.

Thank you.

Humanity is now standing at a crossroads.

London

Can You Hear Me?

My name is Greta Thunberg. I am sixteen years old. I come from Sweden. And I speak on behalf of future generations.

I know many of you don't want to listen to us – you say we are just children. But we're only repeating the message of the united climate science.

Many of you appear concerned that we are wasting valuable lesson time, but I assure you we will go back to school the moment you start listening to science and give us a future. Is that really too much to ask?

In the year 2030 I will be twenty-six years old. My little sister, Beata, will be twenty-three. Just like many of your own children or grandchildren. That is a great age, we have been told. When you have all of your life ahead of you. But I am not so sure it will be that great for us.

I was fortunate to be born in a time and place where everyone told us to dream big; I could become whatever I wanted to. I could live wherever I wanted to. People like me had everything we needed and more. Things our grandparents could not even dream of. We had everything we could ever wish for and yet now we may have nothing.

Now we probably don't even have a future any more.

Because that future was sold so that a small number of people could make unimaginable amounts of money. It was stolen from us every time you said that the sky was the limit, and that you only live once.

You lied to us. You gave us false hope. You told us that the future was something to look forward to. And the saddest thing is that most children are not even aware of the fate that awaits us. We will not understand it until it's too late. And yet we are the lucky ones. Those who will be affected the hardest are already suffering the consequences. But their voices are not heard.

Is my microphone on? Can you hear me?

Around the year 2030, 10 years 252 days and 10 hours away from now, we will be in a position where we set off an irreversible chain reaction beyond human control, that will most likely lead to the end of our civilization as we know it. That is unless, in that time, permanent and unprecedented changes in all aspects of society have taken place, including a reduction of CO_2 emissions by at least 50 per cent.

And please note that these calculations are

depending on inventions that have not yet been invented at scale, inventions that are supposed to clear the atmosphere of astronomical amounts of carbon dioxide.

Furthermore, these calculations do not include unforeseen tipping points and feedback loops like the extremely powerful methane gas escaping from rapidly thawing Arctic permafrost.

Nor do these scientific calculations include already locked-in warming hidden by toxic air pollution. Nor the aspect of equity – or climate justice – clearly stated throughout the Paris Agreement, which is absolutely necessary to make it work on a global scale.

We must also bear in mind that these are just calculations. Estimations. That means that these 'points of no return' may occur a bit sooner or later than 2030. No one can know for sure. We can, however, be certain that they will occur approximately in these time-frames, because these calculations are not opinions or wild guesses.

These projections are backed up by scientific facts, concluded by all nations through the IPCC. Nearly every single major national scientific body around the world unreservedly supports the work and findings of the IPCC.

Did you hear what I just said? Is my English okay? Is the microphone on? Because I'm beginning to wonder.

During the last six months I have travelled around Europe for hundreds of hours in trains, electric cars and buses, repeating these life-changing words over and over again. But no one seems to be talking about it, and nothing has changed. In fact, the emissions are still rising.

When I have been travelling around to speak in different countries, I am always offered help to write about the specific climate policies in specific countries. But that is not really necessary. Because the basic problem is the same everywhere. And the basic problem is that basically nothing is being done to halt – or even slow – climate and ecological breakdown, despite all the beautiful words and promises.

The UK is, however, very special. Not only for its mind-blowing historical carbon debt, but also for its current, very creative, carbon accounting.

Since 1990 the UK has achieved a 37 per cent reduction of its territorial CO_2 emissions, according to the Global Carbon Project. And that does sound very impressive. But these numbers do not include emissions from aviation, shipping and those associated with imports and exports. If these numbers are included the reduction is around 10 per cent since 1990 – or an average of 0.4 per cent a year, according to Tyndall Manchester.

And the main reason for this reduction is not a consequence of climate policies, but rather a 2001 EU directive on air quality that essentially forced the UK to close down its very old and extremely dirty coal power plants and replace them with less dirty gas power stations. And switching from one disastrous energy source to a slightly less disastrous one will of course result in a lowering of emissions.

But perhaps the most dangerous misconception about the climate crisis is that we have to 'lower' our emissions. Because that is far from enough. Our emissions have to stop if we are to stay below 1.5–2°C of warming. The 'lowering of emissions' is of course necessary, but it is only the beginning of a fast process that must lead to a stop within a couple of decades, or less. And by 'stop' I mean net zero – and then quickly on to negative figures. That rules out most of today's politics.

The fact that we are speaking of 'lowering' instead of 'stopping' emissions is perhaps the greatest force behind the continuing business-as-usual. The UK's active, current support of new exploitation of fossil fuels – for example, the UK shale-gas fracking industry, the expansion of its North Sea oil and gas fields, the expansion of airports as well as the planning permission for a brand new coal mine – is beyond absurd.

This ongoing irresponsible behaviour will no doubt be remembered in history as one of the greatest failures of humankind.

People always tell me and the other millions of school-strikers that we should be proud of ourselves for what we have accomplished. But the only thing that we need to look at is the emission curve. And I'm sorry, but it's still rising. That curve is the only thing we should look at.

Every time we make a decision we should ask ourselves: how will this decision affect that curve? We should no longer measure our wealth and success in the graph that shows economic growth, but in the curve that shows the emissions of greenhouse gases. We should no longer only ask: 'Have we got enough money to go through with this?' but also: 'Have we got enough of the carbon budget to spare to go through with this?' That should and must become the centre of our new currency.

Many people say that we don't have any solutions to the climate crisis. And they are right. Because how could we? How do you 'solve' the greatest crisis that humanity has ever faced? How do you 'solve' a war? How do you 'solve' going to the moon for the first time? How do you 'solve' inventing new inventions?

The climate crisis is both the easiest and the hardest issue we have ever faced. The easiest because we know what we must do. We must stop the emissions of greenhouse gases. The hardest because our current economics are still totally dependent on burning fossil fuels, and thereby destroying ecosystems in order to create everlasting economic growth.

'So, exactly how do we solve that?' you ask us – the schoolchildren striking for the climate.

And we say: 'No one knows for sure. But we have to stop burning fossil fuels and restore nature and many other things that we may not have quite figured out yet.'

Then you say: 'That's not an answer!'

So we say: 'We have to start treating the crisis like a crisis – and act even if we don't have all the solutions.'

'That's still not an answer,' you say.

Then we start talking about a circular economy and rewilding nature and the need for a just transition. Then you don't understand what we are talking about.

We say that all those solutions needed are not known to anyone and therefore we must unite behind the science and find them together along the way. But you do not listen to that. Because those answers are for solving a crisis that most of you don't even fully understand. Or don't want to understand.

You don't listen to the science because you are only interested in solutions that will enable you to carry on like before. Like now. And those answers don't exist any more. Because you did not act in time.

Avoiding climate breakdown will require cathedral thinking. We must lay the foundation while we may not know exactly how to build the ceiling.

Sometimes we just simply have to find a way. The moment we decide to fulfil something, we can do anything. And I'm sure that the moment we start behaving as if we were in an emergency, we can avoid climate and ecological catastrophe. Humans are very adaptable: we can still fix this. But the opportunity to do so will not last for long. We must start today. We have no more excuses.

We children are not sacrificing our education and our childhood for you to tell us what you consider is politically possible in the society that you have created. We have not taken to the streets for you to take selfies with us, and tell us that you really admire what we do.

We children are doing this to wake the adults up. We children are doing this for you to put your differences aside and start acting as you would in a crisis. We children are doing this because we want our hopes and dreams back.

I hope my microphone was on. I hope you could all hear me.

We children are doing this because we want our hopes and dreams back.

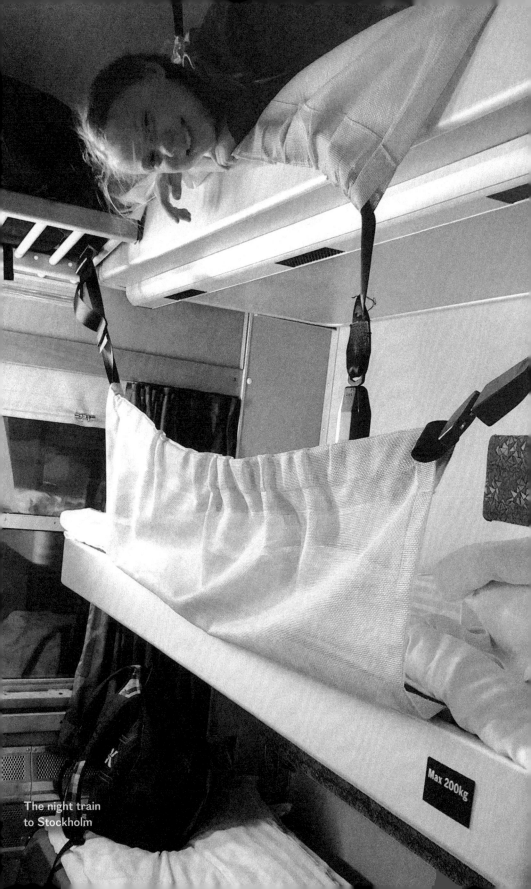

The night train
to Stockholm

May
2019

Vienna

Stockholm

Copenhagen

Vienna

The Easiest Solution is Right in Front of You

My name is Greta Thunberg. I am a climate activist from Sweden and for the last nine months I have been school-striking for the climate every Friday in front of the Swedish Parliament.

We need to change the way we treat the climate crisis.

We need to change the way we speak about the climate crisis.

And we need to call it what it is. An emergency.

I am certain that most of us in here today are generally aware of the situation. But my number one experience during these last nine months is that people in general do not have a clue.

Many of us know something is wrong, that the planet is warming because of increased greenhouse gases, but we don't know the full consequences of that. The vast majority of us know much less than we think.

And this should not be a surprise.

We have never been shown the graphs which show how much the CO_2 emissions must be reduced for us to stay below the 1.5°C limit.

We have never been told the meaning of the aspect of equity in the Paris Agreement – and why it's so important. We have never been taught about feedback loops or tipping points – or what the runaway greenhouse effect is.

Most of us don't know almost any of the basic facts.

Because how could we? We have not been told. Or more importantly, we have never been told by the right people.

We are *Homo sapiens sapiens*. Of the family Hominidae. Of the order Primates. Of the class Mammalia. Of the kingdom Animalia. We are a part of nature. We are social animals. We are naturally drawn to our leaders.

During the last months millions of schoolchildren have been school-striking for the climate and creating lots of attention for the climate crisis. But we children are not leaders. Nor are the scientists, unfortunately. But many of you here today are. Presidents, celebrities, politicians, CEOs and journalists. People listen to you. They are influenced by you. They follow you. And therefore you have an enormous responsibility. And let's be honest. This is a responsibility that most of you have failed to take.

You cannot rely on people reading between the lines or searching for the information themselves. To read through the latest IPCC report, track the Keeling Curve or keep tabs on the world's rapidly disappearing carbon budget. You have to explain that to us, repeatedly. No matter how uncomfortable or unprofitable that may be.

And yes, a transformed world will include lots of benefits. But you have to understand. This is not primarily an opportunity to create new green jobs, new businesses or green economic growth. This is above all an emergency, and not just any emergency. This is the biggest crisis humanity has ever faced. This is not something you can like on Facebook.

When I first heard about the climate and ecological breakdown I actually didn't believe that it could be happening.

Because how could it be? How could we be facing an existential crisis that would threaten our very survival, and yet that wasn't our first priority?

If there really was a crisis this big, then we would rarely talk about anything else. As soon as you turned on the TV, everything would be about that. Headlines, radio, newspapers. You would almost never read or hear about anything else.

And the politicians would surely have done what was needed by now, wouldn't they?

They would hold crisis meetings all the time, declare climate emergencies everywhere and spend all their waking hours handling the situation and informing people what was going on.

But it never was like that. The climate crisis was treated just like any other issue, or even less than that. Every time you heard a politician speak about this they never talked with urgency. According to them there were always countless new technologies and simple solutions that, when put in place, would solve everything.

Politicians one second say that 'climate change is the most important topic, we are going to do everything we can to stop it'. And the next second they want to expand airports, build new coal power plants and motorways and then they fly off in a private jet to attend a meeting on the other side of the world.

That is not how you act in a crisis. And humans are social animals, we can't get away from that fact. And as long as you, the leaders, act like everything is fine and you have things under control, people won't understand that we are in an emergency.

You can't only keep talking about specific isolated solutions to specific isolated problems. We need to see the full picture.

If you say that we can 'solve' this crisis just by maybe increasing or lowering some taxes, phasing out coal in ten or fifteen years, putting up solar panels on new buildings or manufacturing more electric cars, then people will think we can 'solve' this crisis with a few political reforms, without anyone making a real effort.

And that is very dangerous. Because specific iso-lated solutions are no longer enough, and you know this. We now need to change practically everything. We now need a whole new way of thinking.

I know you are desperate for hope and solutions. But the biggest source of hope and the easiest solu-tion is right in front of you, and it has been all along. And it is us people, and the fact that we don't know.

We humans are not stupid. We are not ruin-ing the biosphere and future living conditions for all species because we are evil. We are simply not aware. But once we understand, once we realize the situation, then we act, we change. Humans are very adaptable.

So instead of only being obsessed with finding sol-utions to a problem that most of us do not even know exists, you must also focus on informing us about the actual problem.

We must acknowledge that we do not have the situation under control and that we don't have all the solutions yet.

We must admit that we are losing this battle.

We must stop playing with words and numbers.

Because we no longer have time for that, and in the words of author Alex Steffen, 'Winning slowly is the same thing as losing.'

The longer we wait the harder it will be to turn this around. So let's not wait any longer. Let's start acting.

For too long the people in power have gotten away with basically not doing anything to stop climate and ecological breakdown. They have gotten away with stealing our future and selling it for profit.

But we young people are waking up.

And we promise we will not let you get away with it any more.

The longer we wait the harder it will be to turn this around.

Vienna

July
2019

Berlin

Greta Thunberg

You Can't Simply Make Up Your Own Facts

I have some good and some bad news regarding the climate emergency. I will start with the good news. The world – as a small number people have been saying lately – will not end in eleven years.

The bad news however is that around the year 2030, if we continue with business as usual, we will likely be in a position where we may pass a number of tipping points. And then we might no longer be able to undo the irreversible climate breakdown.

A lot of people, a lot of politicians, business leaders, journalists say that they don't 'agree' with what we children are saying. They say we are exaggerating, that we are alarmists.

To answer this I would like to refer to page 108, chapter 2, in the latest IPCC report.

There you will find all our 'opinions' summarized.

Because there you find our remaining carbon-dioxide budget.

Right there it says that if we are to have a 67 per cent chance of limiting the global temperature rise to below 1.5°C, we had, on 1 January 2018, 420 gigatonnes of carbon dioxide left in our CO_2 budget. And of course that number is much lower today. We emit about 42 gigatonnes of CO_2 every year.

At current emissions levels, that remaining budget is gone within roughly 8.5 years.

These numbers are as real as it gets. Though a great number of scientists suggest they are too generous, these are the ones that have been accepted by all nations through the IPCC.

And not once, not one single time, have I heard any politician, journalist or business leader even mention these numbers. It is almost like you don't even know they exist, as if you haven't even read the latest IPCC report, on which the future of our civilization is depending.

Or maybe you are simply not mature enough to tell it like it is.

Because even that burden you leave to us children.

We become the bad guys who have to tell people these uncomfortable things, because no one else wants to, or dares to.

And just for quoting and acting on these numbers – these scientific facts – we receive unimaginable amounts of hate and threats. We are being mocked and lied about by elected officials, members of parliaments, business leaders, journalists.

What I really would like to ask all of those who question our so-called 'opinions', or think that we are extreme: Do you have a different budget for at least a reasonable chance of staying below 1.5°C of warming? Is there another intergovernmental panel on climate change? Is there a secret Paris agreement that we don't know about? One that does not include the aspect of equity?

Because these are the numbers that count. This is the current best available science. You can't simply make up your own facts, just because you don't like what you hear. There is no middle ground when it comes to the climate and ecological emergency.

Of course you could argue that we should go for a more risky pathway, such as the alternative of 580 gigatonnes of CO_2 from 1 January 2018, which gives us a 50:50 chance of limiting the global temperature rise to below 1.5°C. That amount of carbon dioxide will run out in about twelve years of current business as usual. But why should we do that? Why should we accept taking that risk? Leaving the future living conditions for humankind to a 50:50 flip of a coin?

Four hundred and twenty gigatonnes left of CO_2 to emit. And now that number is down to less than 360 gigatonnes.

And please note that these figures are global and therefore do not say anything about the aspect of equity, clearly stated throughout the Paris Agreement, which is absolutely necessary to make it work on a global scale. That means that richer countries need to get down to zero emissions faster – so that people in poorer parts of the world can heighten their standard of living by building some of the infrastructure that we have already built. Such as roads, hospitals, electricity, schools and providing clean drinking water.

And because you have ignored these facts, because you and pretty much all of the media, to this very minute keep ignoring them – people do not know what is going on.

If you respect the science, if you understand the science, then this is it. Four hundred and twenty gigatonnes of CO_2 left to emit on 1 January 2018 to have a 67 per cent chance of staying below a 1.5°C global temperature rise, according to the IPCC.

In the Paris Agreement, we have only signed up for staying below 1.5–2°C of temperature rise. And that of course gives us a bigger remaining carbon-dioxide budget. But the latest IPCC report shows that aiming instead for below 1.5°C would significantly reduce the climate impacts, and that would most certainly save countless human lives.

This is what it's all about. This is all that we are saying. But I will also tell you this, you cannot solve a crisis without treating it like a crisis, without seeing the full picture. You cannot leave the responsibility to individuals, politicians, the market or other parts of the world to take. This has to include everything and everyone.

Once you realize how painfully small the size of

our remaining carbon-dioxide budget is; once you realize how fast it is disappearing; once you realize that basically nothing is being done about it; and once you realize that almost no one is even aware of the fact that CO_2 budget even exists … then tell me – just exactly what do you do? And how do you do it without sounding alarmist?

That is the question we must ask ourselves, and the people in power.

The science is clear. And all we children are doing is communicating and acting on that united science.

Now political leaders in some countries are starting to talk. They are starting to declare climate emergencies and announcing dates for so-called 'climate neutrality'. And declaring a climate emergency is good.

But only setting up these vague distant dates, and saying things which give the impression that things are being done and that action is underway will most likely do more harm than good. Because the changes required are still nowhere in sight. Not in France, not in the EU. Nowhere.

And I believe that the biggest danger is not our inaction. The real danger is when companies and politicians are making it look like real action is happening, when in fact almost nothing is being done, apart from clever accounting and creative PR.

The climate and ecological emergency is right here, right now. But it has only just begun. It will get worse.

Four hundred and twenty gigatonnes of CO_2 left

to emit on 1 January 2018 to have a 67 per cent chance of staying below a 1.5°C global temperature rise.

And now that figure is already down to less than 360 gigatonnes.

At current emission levels, that remaining budget is gone within roughly 8.5 years.

In fact, since I started this speech the world has emitted about 800,000 tonnes of CO_2.

And if anyone still has excuses – not to listen, not to act, not to care – I ask you once again: Is there another intergovernmental panel on climate change? Is there a secret Paris agreement that we don't know about? One that does not include the aspect of equity? Do you have a different budget for at least a reasonable chance of staying below 1.5°C of global temperature rise?

Some people have chosen not to come here today, some people have chosen not to listen to us. And that is fine, we are after all just children. You don't have to listen to us. But you have to listen to the united science. The scientists. And that is all we ask – unite behind the science.

The science is clear. And all we children are doing is communicating and acting on that united science.

Paris

September
2019

Halfway
across
the
Atlantic
Ocean

Hamburg

Wherever I Go I Seem to be Surrounded by Fairy Tales

My name is Greta Thunberg, I am sixteen years old and I'm from Sweden. I am grateful for being with you here in the USA. A nation that, to many people, is the country of dreams. I also have a dream.

That governments, political parties and corporations grasp the urgency of the climate and ecological crisis and come together despite their differences – as you would in an emergency – and take the measures required to safeguard the conditions for a dignified life for everybody on earth. Because then we millions of school-striking youth could go back to school.

I have a dream that the people in power, as well as the media, start treating this crisis like the existential emergency it is. So that I could go home to my sister and my dogs. Because I miss them.

In fact I have many dreams. But this is the year

2019. This is not the time and place for dreams. This is the time to wake up. This is the moment in history when we need to be wide awake.

And yes, we need dreams, we cannot live without dreams. But there's a time and place for everything. And dreams cannot stand in the way of telling it like it is.

And yet, wherever I go I seem to be surrounded by fairy tales. Business leaders, elected officials all across the political spectrum spending their time making up and telling bedtime stories that soothe us, that make us go back to sleep. These are feel-good stories about how we are going to fix everything. How wonderful everything is going to be when we have 'solved' everything. But the problem we are facing is not that we lack the ability to dream, or to imagine a better world. The problem now is that we need to wake up. It's time to face the reality, the facts, the science.

And the science doesn't mainly speak of 'great opportunities to create the society we always wanted'. It tells of unspoken human sufferings, which will get worse and worse the longer we delay action – unless we start to act now. And yes, of course a sustainable transformed world will include lots of new benefits. But you have to understand. This is not primarily an opportunity to create new green jobs, new businesses or green economic growth. This is above all an emergency, and not just any emergency. This is the biggest crisis humanity has ever faced.

And we need to treat it accordingly. So that people

can understand and grasp the urgency. Because you cannot solve a crisis without treating it as one. Stop telling people that everything will be fine when in fact, as it looks now, it won't be very 'fine'. This is not something you can package and sell or 'like' on social media.

Stop pretending that you, your business idea, your political party or plan will solve everything. We must realize that we don't have all the solutions yet. Far from it. Unless those solutions mean that we simply stop doing certain things.

Changing one disastrous energy source for a slightly less disastrous one is not progress.

Exporting our emissions overseas is not reducing our emissions.

Creative accounting will not help us. In fact, it's the very heart of the problem.

Some of you may have heard that we have twelve years as from 1 January 2018 to cut our emissions of carbon dioxide in half. But I guess that hardly any of you have heard that that is for a 50 per cent chance of staying below a 1.5°C global temperature rise above pre-industrial levels. A 50 per cent chance.

And these current, best available scientific calculations do not include non-linear tipping points as well as most unforeseen feedback loops like the extremely powerful methane gas escaping from rapidly thawing Arctic permafrost. Or already locked-in warming hidden by toxic air pollution. Or the aspect of equity, climate justice.

So a 50 per cent chance – a statistical flip of a coin

– will most definitely not be enough. That would be impossible to morally defend.

Would any one of you step onto a plane if you knew it had more than a 50 per cent chance of crashing?

More to the point: would you put your children on that flight?

And why is it so important to stay below the 1.5°C limit? Because that is what the united science calls for, to avoid destabilizing the climate, so that we stay clear of setting off an irreversible chain reaction beyond human control. Even at 1°C of warming we are seeing an unacceptable loss of life and livelihoods.

So where do we begin?

Well, I would suggest that we start looking at chapter 2, on page 108 of the SR15 IPCC report that came out last year.

Right there it says that if we are to have a 67 per cent chance of limiting the global temperature rise to below 1.5°C, we had, on 1 January 2018, about 420 gigatonnes of CO_2 left to emit in the carbon-dioxide budget. And of course that number is much lower today. As we emit about 42 gigatonnes of CO_2 every year, if you include land use.

With today's emission levels, that remaining budget is gone within less than 8.5 years.

These numbers are not my opinions. They aren't anyone's opinions or political views. This is the current best available science. Though a great number of scientists suggest even these figures are too mod-erate, these are the ones that have been accepted by all nations through the IPCC.

And please note that these figures are global and therefore do not say anything about the aspect of equity, clearly stated throughout the Paris Agreement. Which is absolutely necessary to make it work on a global scale. That means that richer countries need to do their fair share and get down to zero emissions much faster, so that people in poorer countries can heighten their standard of living, by building some of the infrastructure that we have already built. Such as roads, hospitals, schools, clean drinking water and electricity.

The USA is the biggest carbon polluter in history. It is also the world's number one producer of oil. And yet, you are also the only nation in the world that has signalled your strong intention to leave the Paris Agreement.

Because, quote, 'It was a bad deal for the USA.'

Four hundred and twenty gigatonnes of CO_2 left to emit on 1 January 2018 to have a 67 per cent chance of staying below a 1.5°C global temperature rise.

Now that figure is already down to less than 360 gigatonnes.

These numbers are very uncomfortable. But people have the right to know. And the vast majority of us have no idea these numbers even exist. In fact not even the journalists that I meet seem to know that they even exist.

Not to mention the politicians.

And yet they all seem so certain that their political plan will solve the entire crisis.

But how can we solve a problem that we don't even fully understand? How can we leave out the full picture and the current best available science?

I believe there is a huge danger in doing so. And no matter how political the background to this crisis may be, we must not allow this to continue to be a partisan political question. The climate and ecological crisis is beyond party politics. And our main enemy right now is not our political opponents. Our main enemy now is physics. And we cannot make 'deals' with physics.

Everybody says that making sacrifices for the survival of the biosphere – and to secure the living conditions for future and present generations – is an impossible thing to do.

Americans have indeed made great sacrifices to overcome terrible odds before.

Think of the brave soldiers that rushed ashore in that first wave on Omaha Beach on D-Day.

Think of Martin Luther King and the 600 other civil rights leaders who risked everything to march from Selma to Montgomery.

Think of President John F. Kennedy announcing in 1962 that America would 'choose to go to the moon in this decade and do the other things, not because they are easy, but because they are hard'.

Perhaps it is impossible.

But looking at those numbers – looking at the

current best available science signed by every nation – then I think that is precisely what we are up against.

But you must not spend all of your time dreaming, or see this as some political fight to win.

And you must not gamble your children's future on the flip of a coin.

Instead, you must unite behind the science.

You must take action.

You must do the impossible.

Because giving up can never ever be an option.

You cannot solve a crisis without treating it as one.

Cape Town

Paris

Durban

London

Sydney

Berlin

Mumbai

FRIDAYS
OR FUTURE
AFGHANISTAN

Osa Peninsula

♥ 4.3m

💬 134k

This is all wrong.

The World is Waking Up

This is all wrong. I shouldn't be standing here.

I should be back in school on the other side of the ocean. Yet you all come to us young people for hope? How dare you!

You have taken away my dreams and my childhood with your empty words. And yet I'm one of the lucky ones.

People are suffering. People are dying. Entire ecosystems are collapsing. We are in the beginning of a mass extinction. And all you can talk about is money and fairy tales of eternal economic growth. How dare you!

For more than thirty years the science has been crystal clear. How dare you continue to look away, and come here saying that you are doing enough.

When the politics and solutions needed are still nowhere in sight.

You say you 'hear' us and that you understand the urgency. But no matter how sad and angry I am, I don't want to believe that. Because if you fully understood the situation and still kept on failing to act, then you would be evil.

And I refuse to believe that.

The popular idea of cutting our emissions in half in ten years only gives us a 50 per cent chance of staying below 1.5°C and the risk of setting off irreversible chain reactions beyond human control.

Fifty per cent may be acceptable to you.

But since those numbers don't include tipping points, most feedback loops, additional warming hidden by toxic air pollution, nor the aspect of equity, then a 50 per cent risk is simply not acceptable to us, we who have to live with the consequences. We do not accept these odds.

To have a 67 per cent chance of staying below a 1.5°C global temperature rise, the best odds given by the IPCC, the world had 420 gigatonnes of CO_2 left to emit back on 1 January 2018.

Today that figure is already down to less than 350 gigatonnes. How dare you pretend that this can be solved with business as usual and some technical solutions!

With today's emission levels, that remaining CO_2 budget will be entirely gone within less than 8.5 years.

There will not be any solutions or plans presented in line with these figures today. Because these numbers are too uncomfortable. And you are still not mature enough to tell it like it is.

Your generation is failing us. But the young people are starting to understand your betrayal. The eyes of all future generations are upon you.

And if you choose to fail us I say we will never forgive you.

We will not let you get away with this. Right here, right now is where we draw the line.

The world is waking up.

And change is coming, whether you like it or not.

If you fully understood the situation and still kept on failing to act, then you would be evil. And I refuse to believe that.

New York

We are the Change and Change is Coming

Bonjour Montréal! Je suis très heureuse d'être ici au Canada au Québec! Ça me rappelle la maison. Merci! They say we are 500,000 people marching here in Montreal today!

It's great to be in Canada. It's a bit like coming home. I mean, you are so similar to Sweden, where I'm from.

You have moose and we have moose. You have cold winters and lots of snow and pine trees. And we have cold winters and lots of snow and pine trees.

You have the caribou and we have reindeer. You play ice hockey and we play ice hockey.

You have maple syrup and we have . . . well . . . forget about that one.

You are a nation that allegedly is a climate leader. And Sweden is also a nation that is allegedly a climate leader. And in both cases it sadly means absolutely nothing. Because in both cases it's just empty words. And the politics needed is still nowhere in sight.

So we are basically the same!

Last week well over 4 million people in over 170 countries striked for the climate.

We marched for a living planet and a safe future for everyone.

We spoke the science and demanded that the people in power would listen to and act on the science.

But our political leaders didn't listen.

This week world leaders gathered in New York for the UN Climate Action Summit. They disappointed us once again with empty words and insufficient action. We told them to unite behind the science. But they didn't listen.

So today we are millions around the world striking and marching again. And we will keep on doing it until they listen. If the people in power won't take responsibility, then we will. It shouldn't be up to us, but somebody needs to do it.

They say we shouldn't worry, that we should look forward to a bright future. But they forget that if they would have done their job, we wouldn't need to worry. If they had started in time then this crisis would not be the crisis it is today. And we promise: once they start to do their job and take responsibility, we will stop worrying and go back to school, go back to work. And once again, we are not communicating our opinions or any political views. The climate and ecological crisis is beyond party politics. We are communicating the current best available science.

To some people – particularly those who in many ways have created this crisis – that science is far too uncomfortable to address. But we who will have to live with the consequences – and indeed those who are living with the climate and ecological crisis already – don't have a choice. To stay below 1.5°C – and give us a chance to avoid the risk of setting off irreversible chain reactions beyond human control – we must speak the truth and tell it like it is.

In the IPCC's SR15 report that came out last year it says on page 108, chapter 2, that to have a 67 per cent chance of staying below a 1.5°C global temperature rise – the best odds given by the IPCC – the world had 420 gigatonnes of CO_2 left to emit back on 1 January 2018.

Today that figure is already down to less than 350 gigatonnes.

With today's emissions levels, that remaining CO_2 budget will be entirely gone within less than 8.5 years.

And please note that these calculations do not include already locked-in warming hidden by toxic air pollution, non-linear tipping points, most feedback loops, or the aspect of equity, climate justice.

They are also relying on my generation sucking hundreds of billions of tonnes of CO_2 out of the air with technologies that barely exist.

And not once, not one single time, have I heard any

politician, journalist or business leader even mention these numbers.

They say let children be children. We agree, let us be children. Do your part, communicate these kinds of numbers instead of leaving that responsibility to us. Then we can go back to 'being children'.

We are not in school today. We are not at work today. Because this is an emergency. And we will not be bystanders.

Some would say we are wasting lesson time; we say we are changing the world. So that when we are older we will be able to say we did everything we could. And we will never stop doing that. We will never stop fighting for the living planet and for our future.

We will do everything in our power to stop this crisis from getting worse. Even if that means skipping school or work. Because this is more important.

We have been told so many times that there's no point in doing this, that we won't have an impact anyway, that we can't make a difference. I think we have proven that to be wrong by now.

Through history, the most important changes in society have come from the bottom up, from grass-roots. The numbers are still coming in – but it looks like well over 7 million people have joined the Week for Future, the strikes on this and last Friday. That is one of the biggest demonstrations in history. The people have spoken and we will continue to speak until our leaders listen. We are the change and change is coming.

Le changement arrive – si vous l'aimez ou non!

Photograph credits by page number:

Frontispiece, 5 & 8: Stockholm, Sweden © Anders Hellberg

14: London, UK, 31 October 2018. Peter Marshall/Alamy

15: London, UK, 31 October 2018. Guy Bell/Alamy

24: Katowice, Poland, 8 December 2018. Diogo Baptista/Alamy

25: Katowice, Poland, 14 December 2019. Omar Marques/Zuma Press/PA Images

32, 33 & 34: Davos, Switzerland, 25 January 2019. Fabrice Coffrini/AFP/Getty Images

51: Brussels, Belgium, 21 February 2019. Alexandros Michaildis/Shutterstock

52: Brussels, Belgium, 21 February 2019. Yves Herman/Reuters

60: Berlin, Germany, 29 March 2019. Kay Nietfeld/dpa/Alamy

61: Berlin, Germany, 29 March 2019. Michael Kappeler/dpa/Alamy

62: Berlin, Germany, 30 March 2019. Hannibal Hanschke/Reuters pool/dpa/Alamy

68: Strasbourg, France, 16 April 2019. Frederick Florin/AFP/Getty Images

76: London, UK, 21 April 2019. Tolga Akmen/AFP/Getty Images

79: London, UK, 21 April 2019. Phil Clarke Hill/Getty Images

80: London, UK, 23 April 2019. Toby Melville/Reuters

92: *bottom* **Vienna, Austria, 28 May 2019.** Thomas Kronsteiner/Getty Images

94: Vienna, Austria, 28 May 2019. Askin Kiyagan/Anadolou Agency/Getty Images

101: Vienna, Austria 31 May 2019 Leonhard Foeger/Reuters

104: Paris, France, 23 July 2019. Niviere David/ABACA/PA Images

111: Paris, France, 23 July 2019. Lionel Bonaventure/AFP/Getty Images

114: Turin, Italy, 27 September 2019. Giulio Lapone/Getty Images.

115: Hamburg, Germany, 20 September 2019. Joerg Boethling/Alamy

116: Washington DC, USA, 23 September 2019. Alex Wong/Getty Images

124-5: *top left* **Cape Town, South Africa, 20 September 2019.** Nic Bothma/EPA-EFE/Shutterstock
top centre **Paris, France, 20 September 2019.** Daniel Pier/NurPhoto/Getty Images
top right **Mumbai, India, 27 September 2019.** Divyakant Solanki/EPA-EFE/Shutterstock
centre left **Durban, South Africa, 20 September 2019.** Rajesh Jantilal/AFP/Getty Images
centre **London, UK, 20 September 2019.** Kristian Buus/InPictures/Getty Images
centre right **Kabul, Afghanistan, 20 September 2019.** Fridays For Future, Afghanistan
bottom left **Brisbane, Australia, 20 September 2019.** Paul Braven/EPA-EFE/Shutterstock
bottom centre **Berlin, Germany, 20 September 2019.** Axel Schmidt/AFP/Getty
bottom right **Carate, Peninsula Osa, Costa Rica, 20 September 2019.** The Integral
Development Asociation Corcovado Carate

130: *bottom* **New York, USA, 20 September 2019.** Roy Rochlin/Getty Images

131: New York, USA, 20 September 2019. Michael Appleton/Mayoral Photography Office

132. Montreal, Canada, 27 September 2019. Eric Demers/Polaris/eyevine

137: Montreal, Canada, 27 September 2019 © Ben Etienne

All other images courtesy of the author and her family.

www.greenpenguin.co.uk

MiX
Paper from
responsible sources
www.fsc.org **FSC® C018179**

Penguin Random House is committed to a
sustainable future for our business, our readers
and our planet. This book is made from Forest
Stewardship Council® certified paper.

ALLEN LANE
UK | USA | Canada | Ireland | Australia
India | New Zealand | South Africa
Allen Lane is part of the Penguin
Random House group of companies
whose addresses can be found at
global.penguinrandomhouse.com

First published 2019
This illustrated edition published 2019
001

Copyright © Greta Thunberg, 2018–2019,
in agreement with Politiken Literary Agency
The moral right of the author has been asserted
Book design by Jim Stoddart
Cover design by Tom Etherington
Set in Baskerville MT Standard and Figgins Sans
Printed in Great Britain by CPI
A CIP catalogue record for this book
is available from the British Library
ISBN: 978-0-241-45344-5